Thank God I'm Different

Thank God I'm Different
About Asperger's Syndrome

Caroline Bidal

Published by: Caroline Bidal
2014

*Please note that there are no page numbers for the sole purpose in making this book Asperger user friendly.

First Printing: 2014

ISBN 978-1-304-80680-2

Caroline Bidal Publishing

Ste-Adele, Quebec, J8B 2B2

450-745-1577

www.CarolineBidalPublishing.com

Ordering Information:

Special discounts are available on quantity purchases by corporations, associations, educators, and others. For details, contact the publisher at the above listed address.

Dedication

To my wonderful husband and best friend.

Thank you. Without your support and patience, I would have never achieved this book.

Contents

Chapter 1: Been Misunderstood

Autism is a spectrum condition which means that, while people who have autism share certain difficulties, their condition will affect them all in different ways. The main fields of difficulty are interacting and communicating. Feeling overwhelmed in certain situations that don't seem overwhelming to most people, lack of understanding of the unwritten rules of society, over-sensitivity to sound, touch, taste, smells, light, colors, balance and movement. Asperger's syndrome is "autism with language"; the person with Asperger's syndrome did not have a language delay as a child but has social deficits and the perceptual differences of an autistic person on the lower end of the spectrum.

We can often act bizarre or out of standard, behave inappropriately or strangely as it is not always easy for aspies to express their feelings, emotions and needs. We may therefore, appear insensitive because we do not recognize how the other person feels.

People with AS have a totally different point of view of the world than the general public. As is today considered as an autistic variant and part of the development impairment disorder group. Often once the patient is diagnosed with AS, they often also diagnose with Attention deficit disorder.

Here are the principal symptoms of AS:
- Lack of empathy
- Unilateral social interaction, can be naïve and inappropriate
- Retained or limited capacity in establishing friendships
- Repeated verbal skills, pendant language skills
- Intense preoccupationnal focus on certain subjects
- Clumsiness and lack of coordination in movements, bizarre attitudes.

Dr. Hans Asperger defined Asperger's syndrome approximately 50 years ago. The syndrome was characterized as a low empathy level, a very weak sense of making friends, direct or linear conversations, polarization of interest and a poor sense and lack of motor skills.

As an aspie, we have to learn to engage in the world as hard working members of society. We may seem weird or bizarre to people but we have wonderful abilities that we must discover within ourselves and propel towards the exterior world.

Aspies, once again, have a lack of social competence, a limited ability to engage in a two way conversation and are often involved in an intense interest in a specific field. All of these traits do not stop us from being aware of our environment.

Aspies do not abby with conventional methods. They have their own unique ways of thinking. They do need help to learn how to acquire social skills needed to properly interact within a group. They can obviously benefit from an enriching adult relationship.

I think in images, in my mind, it is like having photographs flipping one by one. Thinking in abstract ways is not very easy for aspies to do. This can be hard for our surrounding family who cannot picture our thoughts.

We have a wonderful ability to think about things in depth, for long periods of time.

There are stages that we go through that makes us realize that we have Asperger's:

- **Awareness**: We discover and read about Asperger's syndrome but the idea of it has not hit home yet.

- **Knowing**: The irreversible understanding that Asperger is what you have. It starts to sink in.

- **Validation**: Asperger's syndrome explains so much in our lives that seemed that it had no rhythm or reason. This will be a series of events that will keep on happening.

- **Relief**: We finally get a diagnosis and everything explains itself but we can tell other people that do not seem to be carrying it.

- **Worry**: How will this affect my life in its entire dimension? How will this affect my potential?

- **Anger**: For all that we went through, misdiagnosis, blame and misunderstanding but now we can get along with our lives.

- **Acceptance and learning to thrive**: We become aware of our great gifts and deficits that we learn to exploit and use in a wise matter.

Chapter 2: Sensory overload & Meltdowns

Meltdowns are badly seen from the eyes of the common. Close your eyes, imagine being in a room with 20 people, they are all talking at the same time, divided into small conversation groups. The music is very loud playing this high pitched lounge music that just seems to be getting louder and louder. You can hear glasses being clanged together as they propose a toast. Cellular phones going off, you can hear car alarms outside, the wind hitting the side of the building, the bathroom door closing with the rumble of the toilet flushing, mood lighting and party lights flashing away, the heat that intensifies the bodily odors of all the people gathered close together. We can be set off by loud social situations, traffic, standing in line, strong perfumes, fluorescent light and the noise they emit. Loud music in shopping malls, sirens, fire alarms, loud power tools, commercials on TV that get louder once they are playing between your favorite shows.

Knowing what set us off and realizing what triggers them is just the beginning what it feels like and why we act upon them.

Sensory processing disorder is quite a separate syndrome and is not part of the many triggers of a meltdown. It is part of the autistic package. Most of us will have it to some degree or may not have it at all.

All this sensory mix down is so overwhelming that we cannot deal with the pain and I mean physical pain of the entire sensory overload that our brains analyze at a faster rate than the normal. We just want to get away so all this craziness stops and we can be back to a stable state of calmness and quietness. If we cannot escape it, we will crash and break down into a temper tantrum, screaming and yelling, recollecting ourselves in a corner and swinging our bodies or flapping our hands to relieve the intensity of the pain we suffer from.

When I attend social circles and try to fit in even trying to act normal tired me out and drains all of my energies. We do not like to be touched or hugged by other people because it physically hurts and our bubble is sacred. I have this amazing cover in my bed that weighs 30 lbs and it helps me find the sleep I need by allowing this constant even spread pressure across my body that soothes and calms me down.

We do not like changes in our routines or by an unexpected situation, this will through us off and make us very anxious to the point of inducing a meltdown. We find it very difficult to cope or more like adapt to new or unfamiliar situations or when a set of important rules and guidelines are broken.

Sights

Flickering Christmas lights on trees on the public plazas is beautiful to many but for aspies that can just set us off and be quite disturbing. The flashing hues hit our eyes and it is so painful that we will want to hide and run.

When I drive at night to go to work, I cannot stand high beams from the upcoming cars ahead on the opposite lane. The intensity of their lights varies from if it is an old or a more recent car. Most people will take them off as you cross each other but I dread the ones who do not. The intensity of the light is so intense that I almost drove off the road. I feel like a deer trying to cross the road but get stunned by the big lights ahead, frightened and somewhat experiencing fear of the unknown result of such a surge of energy.

It makes me upset to the point that I talk out loud about how they should turn those damn lights down and that it is such a lack of respect for the other drivers.

I was in a small town about 30 minutes from my house the other day, driving along the main street when up ahead I saw a flashing square not only flashing but more like a strobe. I calculated quickly that they had time to flash ones every 2 seconds interval. All my attention diverted towards that light and I was trying so hard to look elsewhere to stop the thumping pain that it was causing through my eyes into my frontal lobe.

Sight overloads does not happen just with the essence of lighting, it also happen with the movement of people, crowds as they are moving about. These are all sensory issues that can be extremely difficult to deal with all at ones.

We package in so much information in our brain quickly that our brains overload. This can make us very anxious at times like when u see the classic autistic temper tantrums in line at the cash. Shopping malls are hell on earth for aspies of all ages. We take in too much information for our brains to handle.

I shop in small local stores where I am use to their setup and know where everything is therefore I can get around quickly and get what I need with no fuss. The staff is usually the same so familiar faces help break the social barriers. If I have to go to a large warehouse type store, I go during the week off rush hours and always to the same one so I know where the items are that I have on my shopping list. I never go alone neither, my husband also an aspie, come along all the time, we encourage ourselves to be able to survive the frenzy.

Christmas is on its way, I refuse to even step or go near a store during busy hours. I live up in the mountains, in a village where there are small privately owned stores none of these large chained stores for a reason. I do not have to get loads of Christmas gifts therefore I try to do the shopping online or off hours. Internet shopping is great for that, no malls, no crowds, no anxiety only when u get the bill and most of all delivered to your doorstep. Obviously, you have to be prepared for the delivery personnel but it's a lot better than standing in line at a major retail store.

Sounds

To follow a conversation properly bringing down the surrounding sounds are essential. When I go over to my folk's house, I immediately either turn down or completely shut off the television or music. My folks are use to this so it does not insult them.

In the car with my husband, I cannot follow a conversation if the radio is blasting away or if I'm driving I just do not listen to any music.

At bedtime, sounds can be a big challenge. I usually make sure that nothing is on and if I need so earplugs are a good option. I take medication to induce sleep but it's a short term alternative. I have tried hypnotherapy and that is a success. I also sleep with a 30 lbs blanket especially made to help soothe anxieties and help with sleep. My three dogs usually cuddle underneath the blanket. They benefit from the pressure of the blanket.

Smells

I get really noxious when I smell garlic in a supermarkets meat department when they are preparing Italian sausages. The smell is so intense to me that I literally say out loud oh boy it stinks here at the customer choosing his favorite sausages in the display.

Intense smells can even lead to an allergic outburst of sniffles and itchy eyes.

Sensory Tools

Sensory Tools are amazing to cope with sensory issues or to prevent meltdowns before they happen. A pair of ear plugs really comfy clothes without any elastic around our sleeves, a stress squishy ball on hand, relaxing music on an IPod, hats and sunglasses. A gluten free diet without ingesting any chemicals, eating bio fruits and vegetables, avoid sugars and processed foods.

Sensory rooms are also popular among aspies; it helps regulate meltdowns before they happen. Some rooms have special lights, smells, sounds, textures and temperature. There are also sensory deprivation rooms which are limited in their output in emotions and sensory issues we all need time out from noises or smells.

Meltdowns can be caused by simply been bombarded with sensory stimulations that are partially overwhelming and incomprehensible emotional distress. Lights, sounds, odors, and taste involving textures can be overwhelming and be part of the many triggers of a major meltdown.

Finally, stress and sensory overloads needs to be looked at in a total bundle. Environmental sensitivity will exist for most of us but our reactions will be different and can be managed through taking good care of ourselves.

Chapter 3: Conversations & social interactions

Being asperger means what to the world. But what does the world mean to an asperger. It means being different not less intelligent. Actually most aspies have above average IQs. We have great hobbies or interests, we hyper focus on specific details, which usually become career oriented.

We lack social skills from an early age; we do not like to be surrounded by large crowds, to many sensory issues to deal with all at ones. What seems overwhelming to an aspie will not seem overwhelming to a normal person that would be put in the same social situation. We might act weird or misconduct behaviors that could offend people in general but it's all part of the lack of knowledge and misinterpretation of social situations.

We get very anxious to certain situations because we lack the notion of the consequences of our actions. It is very hard to take a decision when you cannot properly analyze the outcome of the decision that you decided to act upon. Who to trust, who to turn to, who can we let in our world that will understand and comprehend our dilemmas. Who will not take advantage of our innocents or the fact that we cannot properly understand sarcasm? It takes someone really special to whom we can really trust because it is so difficult to trust people.

People with asperger dislike physical contact. They avoid eye contact simply because they are under pressure and anxious. To some people this may seem as a lack of respect and we have something to hide. We often do not understand the social rules that could be inappropriate towards others. People may stand too close to us or we may start inappropriate subject of conversation.

As a child, I was isolated from others, could not integrate into a social group and was often rejected due to my clumsiness. I use to hide out in the front of my house in a red maple tree. That was my safe place which would help me cut off from all the sensory triggers of my day. I had a major meltdown when my father decided to cut the tree down. I remember manifesting against the idea of him cutting down my sensory tree by climbing in it and yelling and crying while my father was cutting the branches around me.

I was terrified and having a major meltdown just then. I was approximately five years of age.

Sarcasm, jokes, similes or metaphors can be understood in the literal sense. We will be misguided by facial expressions, gestures and different tones of voices which we have major difficulties in acquiring the righteous meaning to them.

Therefore, because of our difficulties with social communications, some aspies may seem argumentative, stubborn or even angry. But will be blunt and rude to transmit their point of honesty about the subject in line. Or on the other hand, they may even be over-compliant and agree to everything the other person is saying, even if it's not true just to fit in within the conversation.

Some people who have autism may not speak at all, or have fairly limited, or sometimes stilted, speech. Others will have good language skills but may find it very hard to fully understand the give-and-take nature of conversation. The person might repeat what the other person was saying.

Aspies have a very hard time with understanding or interpret others thoughts, feelings and actions, and how to predict what will or what could happen next. They do not for see the consequences of their actions or understand the concept of danger or bad decision making.

We have intense special interests that we hyper focus on and we love to talk about even in situations that are out of place. It can disengage a conversation because the other person listening will lose interest since it is not the core subject to start off with.

During a conversation with someone who has Asperger syndrome, we may speak properly and clearly but lack full understanding of the conversation. Our apparent independence can mask our isolation from the world and our social disabilities.

Here are a few tricks and tips that will help an Asperger in a social situation.

- Always make sure that the person understands the topic and explain what you are about to engage in. Stay calm, patient and tolerant. Speak clearly in a consistent tone of voice and allow the person some time to process what you are saying. Ask straight forward, clear unambiguous questions that are always in relation to the conversation.

- Avoid using sarcasm, irony, metaphors and similes because they can be taken literally and be misunderstood by the Aspie. Do not engage in any physical contact and remember that eye contact will not be possible and it does not imply shiftiness or disrespectfulness.

- In social situations, Aspies are often misunderstood by other people. Some of us want relationships and some of us rather remain celibate. We can be perceived as been cold or self-centered and we fixate and usually can't understand gender roles.

- We get really excited if we start talking about our favorite and hyper focused topics. On the other hand, we can be shy and even become mute during conversations. We tend to shut down if overloaded in social situations but if we are exposed to small doses we can defiantly appear skilled and outgoing during conversations. It may even look like a performance or an act.

- We usually don't have close friends around us; we prefer to spend our time with our children or partners. Some of us really enjoy the company of animals instead of investing in a friendship. We are very confident in a controlled environment and are very attached to spending time at home.

I love to dress very comfortable and I lack fashion sense. Spending hours in front of the mirror, applying make-up or doing my hair in fancy ways are not even a thought.

As we often socially isolate ourselves, having total self awareness of our bodies, awkwardness, tics, lack of reciprocal social skills, can defiantly limit our choices in finding a suitable partner or friendships.

We are usually more aware that in beauty products that are pressured on us by the multiple companies, we understand that they may contain ingredients that might be harmful in the long run. So we decide to abstain ourselves from their uses. This might be regarded as us being ungroomed and inattentive to our hygiene. If we do not use for example anti-perspirant because we know that they contain ingredients that are proven to be harmful to us, we often turn around and prefer using natural products to keep ourselves as healthy as possible.

People with Asperger's have a fight-to-flight reaction to all social contact. We always want to fill our minds with knowledge the way others want to fill their shopping cart with groceries. Information replaces confusion and chaos, which we experience with others while we interact with them in social situations. It gives us the choice on how much we want to let in our brains and we can keep a complete control over it, unlike dealing with people, who are always uncontrollable and unpredictable. Information defiantly fills a big void while we might be searching for our identities.

Chapter 4: Interaction with animals

Animals have always been great companions for humans from the start of time. Animals do not judge, criticize or reject us. They are simple when it comes to emotions, their body language is easy to determine and they make us get up and do exercise even in our worst state. They find a way to make us smile even when we had a really rough day.

Asperger's Syndrome is an <u>autistic spectrum disorder ('ASD')</u>, a 'hidden disability', which makes it harder to make sense of the world, process information and interact and communicate with others, even though we are articulate and intelligent people. This creates a huge amount of pressure for us when trying to fit in with our peers and relate to them, causing us to experience a high level of social anxiety and, in turn, emotional exhaustion.

Being able to 'talk' to our pets was and is very therapeutic and calming for Aspies, helping us a great deal with the particular kind of stress. We relate better to our pets than to other people because we can sense their needs and understand them and we don't feel the pressure or stress of human relationships.

The love we have for our pets, the love we receive back from them has been a steadying force of peace and security in our sometimes chaotic 'Aspie' way of life. They provide comfort, love and interaction without the expectation for us to say and do the right things. Nothing is expected of us, there is no pressure to 'fit in', we are not being judged, we can be ourselves, and relate to them in love, by cuddling and showing emotion, where otherwise this can be difficult for Aspies.

There seems to be a unique chemistry between people on the autistic spectrum and animals.

The well-known American <u>Temple Grandin</u> is an autistic woman who was incorrectly diagnosed as brain-damaged and developmentally disabled as a very young child and as such, faced many difficult challenges growing up. However, she went on to become a professor at Colorado State University and is now a leading animal behavior expert and consultant to the livestock industry (*credit: Wikipedia*).

As a truly inspirational autistic activist and best-selling author, she tells of her love for horses and how a completely unexpected turning point in her life changed everything for her: While looking after horses at her sister-in-law's ranch, she began to thrive. Not only that, but she also began to feel a special bond with the cattle also at the ranch, in whose company she felt more peaceful than she did with people.

The extremely important value, then, of animals and pets in the lives of those on the autistic spectrum cannot be denied.

Animals provide unconditional acceptance. The dog is delighted to see you, despite the day's disappointments and exhaustions. The horse seems to understand you and wants to be your company. The cat jumps on your lap and purrs in your presence.

There is a natural affinity between animals and people with autism and AS. Thus, pets and animals in general can be effective and successful substitutes for human friends and a menagerie becomes a substitute "family".

Animals identify with and feel relaxed in the company of a non-predator and pets can be a source of comfort and reassurance. A special interest in and natural understanding of animals can become the basis of a successful career. It has been found that children and adults with autism and AS are more able to perceive and have compassion for the perspective of animals than humans.

For Aspies, animal interactions have so many good benefits, they help reduce anxiety, and they transmit to us this sense of well-being just by a cuddle. They are good listeners and will not judge our meltdowns or odd behaviors.

Zoo therapy is an excellent tool to add to an Aspies daily routine. Just by walking your own dog just for a few minutes can be beneficial since you went out there and got some fresh air. If you do not own or cannot have your own companion, helping out at the local animal shelter, walking the k9s or spending some quality time with the cats can certainly brighten both your days. You will see and feel their appreciation.

Horses

Equine zoo therapy is used in various cases of autistic people. More often used with children, there seems to be a connection established between the autistic rider and the horse itself.

Take a moment, look at a horse straight in the eye and you can see their soul. Horses will let you in and they are just amazing creatures.

They teach us about how to stay safe around them because they are animals and we cannot erase the fact that they do weight 1200 lbs. Autistic children lack the sense of staying safe therefore this can absolutely be a great tool for raising awareness.

They also reduce our level of anxiety from daily life just by brushing them or riding them lightly.

People with Asperger's syndrome are very good with animals because their thinking is more sensory oriented than word based. Everyone who encounters horses on a steady schedule should learn to be more sensory oriented in their relationship with them. Sit down in a quiet place and visualize your horse. Is it lying down or standing up, is it frightened or flighty. Keep in mind that when working with your horse, keep the sensory details in your mind.

Dogs

Dogs are great when it comes for companionship. I always bring my Chihuahua Chico to any social meets I have with the family. He helps me develop conversations and it takes the pressure off about the difficulties in engaging in elaborate conversations. I also have to take him out from time to time so he can do his business. These little moments give me a time out from the turmoil's of overwhelminess. They also love to be walked so they get you out of the house. They need exercise therefore they make you do exercise. They have this unconditional love that never vanishes and they are always willing to get a cuddle or two. When it comes to bedtime, they are ready when you are and will come and snuggle against you under the covers. They are great vacuums when it comes for supper time and cooking.

Dogs have an extremely good sense of knowing if the people who they meet have good vibes or not. This can help you distinguish if that person is right for you or not. Remember if you are looking to adopt a lifetime canine partner be sure to visit your local SPCA or look on petfind.com to find the right partner for you. These dogs are often already well trained and you do not have to go through the whole toilet training which is not always easy to do. Training a dog at an early age without experience can be a disaster if not well supervised by a certified dog trainer. So always keep in mind that there are loads of amazing well trained dogs that are awaiting good forever homes at your local shelter.

Chapter 5: Tips that do help!

We all travel this life with our own self-thought guidelines but we often need to fortify them and earn advice from others around us to help us grow and become stronger.

Here is some advice that was given to me that I would love to share with you.

➤ Families please take the time to educate yourselves on the subject, and educate the people around you.

➤ You are never alone in this world. It might feel that way, but there are internet blog, chat room or even a meeting group in your area.

➤ Love yourselves for who you are, you cannot change yourselves but you can improve.

➤ Do not take the opinions of others in the literal sense; always ask them to speak from a very simple and non ambiguous matter.

➤ Do exercise regularly, improve your diet.

➤ Be very proud that you are different.

➤ Always be yourself and cultivate your talents and transform them into skills that are marketable.

➤ Simplify your life and always be understanding and patient of others.

➤ Be receptive of other people's observations. Don't trust intuition that you do not have.

➤ Be very aware of your sensory overloads so you can manage them well to reduce the meltdowns that they can create.

➤ After a meltdown, try to analyze it to find the triggers that could have caused it. You don't have to find them on your own. Ask for some help.

➤ If you feel uncomfortable filling in a role then do not become that role. Stand out of it.

➤ Select your entourage carefully with people that you can fully trust.

➤ Make life as simple as possible for yourself.

➤ Learn to listen to others in an empathic way and not a sympathetic way. This will take away a lot of the stress of an ongoing conversation filled with other people's emotions.

Chapter 6: Burning social bridges

I remember as a pre-teen as soon as something got too complicated either with social situations, school or even certain hobbies I would leave them. My grandfather believed very much in me and always encouraged me in trying out all kinds of different activities.

He had enrolled me in private piano lessons. I was pretty good but was terrified by the funny somewhat crazy cat woman of a teacher. Then I got to try out Violin lessons which I really enjoyed and participated in a concert put up by the music school. After that I let it go because I got so overwhelmed by the stage tract and the amount of people that were all starring at me on stage that I totally freaked out and never again did I want anything to do with situations like that.

Burning bridges are often the result of a depression meltdown. It is very destructive but on the moment, we feel empowered about what goes on in our lives and we like the control we can have over the situation. We act on the moment but when we are cooled headed then we realize that we still wanted to continue that certain hobby.

Chapter 7: Imagination and self-taught skills

I love information or discovering new things about nature. Information gives our thoughts an anchor; it gives us an identity and something we can control.

Being able to self-teach ourselves is a gift in its own matter. The skills levels off as we get older. This early ability to read understand at an early age, give the impression that we as Asperger's have an air of intellectual maturity that often trick our entourage into thinking that we possess emotional maturity also.

Some Aspies are less language oriented and demonstrate certain learning difficulties such as dyslexia. Throughout our lives we often prefer to be self-taught than to attend higher learning in a college or university.

We love to teach and research ourselves on all the topics because we have our own ways of ingesting and understanding the information.

Recently, it was demonstrated that Asperger's have a higher rate of fluid intelligent than non autistic people. Fluid intelligence is the ability to see order in chaos, to draw lines and understand the relationship of unrelated topics. We may not have higher crystallized intelligence but this explains why we just know how to do something like how to solve a complex math problem or build upon an idea.

We often have or fallen in the slow learning categories. This may happen due to the affect of our difficulties in processing sensory issues around us. We may even shut down or have a meltdown under pressure by people around us.

Autism can be overlooked because our usual obsessions can be considered normal things that we do like music, art, and love for animals. It is the intensity that intensifies how much we enjoy them and the passion they inspire us to upgrade them to a whole new level.

Obsessive behaviors that develop into activities are an amazing illustration of the incredible focus that Aspies possess. The teaching communities, the medical groups are starting to figure out that this is an excellent asset that we have developed. This really should not be discouraged; it really should be reinforced as a strong quality.

We have to be careful because when we are in our "zone " we can well forget to go to the toilet, eat food, take breaks between ideas, drink a glass or two of water, take a shower or get a puff of nice fresh air.

We probably will engage in the same activities all along our lives. Making them more and more amazing and interesting somewhat more complex for us. We can try to share them with others but we get so enrolled in them that we focus all our conversations about them and may even bore them easily.

But let's not forget that there are great aspegians out there that have made a very significant difference and have revolutionized our world like Dr. Temple Grandin, Albert Einstein, and Dan Aykroyd.

As an aspie remember that life is about making a contribution and not about trying to fit in and become the most popular. We also have to take control on our activities like when to know to take a healthy break so many more ideas can be developed. This is excellent for your mental and physical health and without them you would not be able to undertake your favorite activities.

You have to keep the machine finely tuned just like a race car. Without that extra important care and maintenance you would not be able to perform like you should. This could cause a meltdown because you will feel tired and overworked.

Why not try to turn a passive activity into an active one. I sometimes take my laptop to the stable to write while my horses are outside in the paddock running and enjoying themselves. This has defiantly inspired me into writing more and more ideas.

As you develop your passion into a healthy one, more passionate you will become about it and this is an excellent tool to help you deal with further social situations that can be difficult.

If you don't have a savant skill you don't have to worry because you can develop one by getting to know yourself better. Praise and encourage yourself when trying out new hobbies. You will unfold and develop that special talent you have been hiding for so long.

Not understanding things the way usual people do is fine in an academic way because we can usually find our own methods, but in social situations, this game is played differently.

Chapter 8: Marriage and relationships

Here are the fourteen practical strategies for facilitating an AS marriage:

1. Pursuing a diagnosis;
2. Accepting the diagnosis;
3. Staying motivated;
4. Understanding how AS impacts the individual;
5. Managing depression, anxiety, obsessive compulsive disorder and attention deficit hyperactivity disorder;
6. Self-exploration and self-awareness;
7. Creating a Relationship Schedule;
8. Meeting each other's sexual needs;
9. Bridging parallel play;
10. Coping with sensory overload and meltdowns;
11. Expanding Theory of Mind;
12. Improving communication;
13. Co-parenting strategies;
14. Managing expectations and suspending judgment.

1. Pursuing a diagnosis

Diagnosis is an important step in starting to work through issues in an AS marriage. Even if the diagnosis isn't formal, but the couple is able to acknowledge the characteristics and traits of AS that might be causing marital discord, it is very helpful tool to lessen or remove the blame, frustration, shame, depression, pain and isolation felt by one or both partners. In some cases, even if the husband refuses to get an evaluation, the wife may be able to use her understanding of his probable AS to reframe her understanding of her husband and change how she relates to him.

A diagnosis of AS can be obtained from a clinician or a psychologist/neuropsychologist,experienced in identifying AS in adults. It is especially helpful if the clinician's procedure includes interviewing the spouse or partner and/or other family members. Diagnosis can also help with finding an appropriate couple's counselor who can work within the AS framework. Many couples report that working with a couple's counselor who is *not* experienced in working with adults with AS can often harm rather than help the AS marriage.

2. Accepting the AS diagnosis

While re-evaluating the relationship in light of the new diagnosis, and striving to achieve acceptance, it is helpful for both partners to continue to seek information about AS, see a clinician experienced with adult AS, and/or join support groups focused on AS marriages or relationships. A detailed understanding of AS—both the challenging and also the positive traits—is important. Individuals with AS can have some highly desirable traits such as loyalty, honesty, intelligence, strong values, flexibility with gender roles, the ability to work hard, generosity, innocence, humor and good looks. Enumerating all the positive and challenging traits of both partners can give the couple a more balanced picture of their marriage.

3. Staying motivated

It is helpful if both partners are motived to address the issues in their marriage and commit to its long-term success. Otherwise, any attempts to improve the marriage may be short-lived.

In some cases, however, the non-AS partner may be depressed, angry, lonely, and disconnected from her AS partner, that salvaging the marriage is not an option. In such a situation, the couple can work with a couple's counselor or mediator towards an amicable divorce (and resolution of co-parenting issues if they have children involved).

4. Understanding how AS impacts the individual

Psycho-education is an important part of sorting out the challenges in an AS marriages. There are many books on AS marriage written from the point of view of the non-AS partner. Reading such personal relationship narratives can help the non-AS partner by validating her experience and feelings within the marriage. Some narratives paint a painfully negative picture; while it may still be helpful to read these accounts, it is good to keep in mind that every marriage and relationship is unique.

Psycho-education can be a lifelong process, because AS is a rather complex. Traits and behaviors evolve and change through the lifespan of each individual. It's helpful to stay motivated to keep learning about one's partner through the lifespan; there is always more to discover about one another. Similarly, neurotypical traits and behaviors are mysterious and surprising to the partner with AS, and merit continued study and attention. It helps to stay motivated to keep learning about one's partner throughout the lifespan; there is always more to discover about one another.

5. Managing depression, anxiety, OCD, and ADHD

People with AS are at increased risk for depression, anxiety, obsessive compulsive disorder (OCD), or attention deficit disorder/attention deficit hyperactivity disorder (ADD/ADHD). Undiagnosed and untreated anxiety is a major problem for individuals with AS, and can lead to a deeper manifestation of the negative AS traits like impulsivity, melt-downs, rage, and withdrawal, all negatively impacting the marriage. It is vital to diagnose and treat depression, anxiety, OCD, or ADD/ADHD either with medications or/and with therapy.

Another helpful form of intervention can be provided by a life coach who specializes in AS, Coaches can help adults with AS resolve practical problems that are draining their emotionally or causing friction with their spouses, such as employment issues, or difficulty with time management, staying organized, or social skills.

Non-AS spouses can often experience their own mental health issues such as anxiety, depression, affective deprivation disorder, and post-traumatic stress disorder, as a result of being in a relationship with an undiagnosed and untreated partner with AS for an extended period of time. In these cases, the non-AS partner should also receive treatment.

6. Self-exploration and self-awareness

In many AS marriages the non-AS partner may be a super nurture, manager, and organizer, who entered the relationship motivated by a desire to help and nurture the partner with AS. Understanding why she chose her partner with AS is an important step toward becoming self-aware and making changes in her own behavior. Their experiences in their family of origin may have led them to seek out a spouse with AS because he felt familiar. Some of the non-AS partners also say that, when they were going through a vulnerable time in their lives, the strong, quiet, gentle, highly intelligent, and loyal presence of the AS partner provided a sense of emotional security.

Another aspect of self-exploration and self-awareness for the non-AS spouse is to rebuild her self-esteem and reintroduce activities and interests into her life that she may have given up in order to shoulder majority of the responsibility for maintaining the household. The non-AS spouse may also need to look for emotional support outside the marriage, so that she is not solely reliant on her husband for emotional fulfillment—as that may not always be a realistic expectation.

7. Creating a Relationship Schedule

An online and/or paper calendar for important weekly, monthly and yearly events such as holidays, birthdays, anniversaries, family visits, and doctors' appointments is a useful tool for any marriage or relationship. In an AS marriage, adding to this calendar quiet time, times for conversation, sex, shared leisure activities, exercise, and meditation/prayer can be very beneficial to keeping the partners connected on a day to day basis. Based on this calendaring system, couples might want to work on a Relationship Schedule for their marriage.

For example, having daily scheduled conversations between the spouses can serve to keep the couple connected and in-sync with each other on a daily basis, despite the challenges and many activities of everyday life. In addition to scheduling conversation time, it can be beneficial to also schedule sex in order to meet the needs of both partners.

8. Meeting each other's sexual needs

Adults with AS tend to either want a lot of sexual activity or too little; so having a discussion on which days and times to have sex eliminates the guess work for both partners. It is helpful for both partners to communicate their sexual needs verbally, in a clear and detailed manner. Putting sex on the Relationship Schedule isn't enough. Neurological differences apart, people have major differences in how much sex they need, how often, and how they want to be intimate with their partners. Some individuals with AS can be very robotic or technically perfect in bed without paying attention to their partner's need for an emotional connection and foreplay before intercourse. Some individuals with AS also don't enjoy sex due to their sensory issues and/or low sex drive.

It is important for the partner with AS to understand that their partner's sexual needs are different than their own, and that both partners need to work at the keeping emotional connection going on a daily basis, both inside and outside the bedroom.

9. Bridging parallel play

Many couples tell us that common interests and activities is what first brought them together: long walks, boat rides, hikes, picnics, dance events and exercise classes, travel. After getting married, however, many of these joint activities tend to fall off the couple's schedule due to life obligations. Many couples in an AS marriage tend to engage in what is known as "parallel play," where one partner engages in a preferred activity or hobby alone, rather than seeking out his or her partner to enjoy these activities together. Individuals with AS struggle with social/communication initiation and reciprocity. A husband with AS can literally go days, weeks, or even months without spending quality time with his non-AS partner, leaving the non-AS partner feeling abandoned, isolated and terribly lonely.

Research has shown that couples that play together stay together. Playing together—participating in joint leisure activities—can help bridge the physical/emotional distance that is oftentimes is characteristic of an AS marriage. Integrating each other back into the activities that both partners enjoy is beneficial. Once the couple works on creating new memories through shared activities and interests, they can then begin to experience more closeness and togetherness.

10. Coping with sensory overload and meltdowns

Individuals with AS oftentimes have sensory issues. That is, one or more of the person's five senses may be either hypersensitive (overly sensitive) or hyposensitive (with low or diminished sensitivity). For some people with AS, a light caress of the skin can feel like burning fire. Fluorescent lighting can induce an immediate migraine. The noise at a train station, or too many people talking at once at a party, can feel like the loud hammering of metal on metal.

Smells at the grocery store can feel nauseating and overwhelming. On the other hand, a hard prick by a needle can have no effect, or, one could have a diminished sense of smell or taste.

A self-aware and motivated adult with AS can succeed in avoiding meltdowns by learning to avoid the triggers and recognize the early warning signs of stress and sensory overload. Developing strategies to act in response to the early manifestations of an oncoming meltdown can help the spouse with AS.

The non-AS spouse can assist her spouse with AS on his journey to self-awareness. For example, the non-AS partner may be able to bring attention to the AS spouse's rising stress level, and suggest that each of them take some time alone to alleviate some of the stress and overstimulation.

11. Expanding Theory of Mind

Individuals with AS tend to have weak Theory of Mind, meaning a relatively limited ability to "read" another person's thoughts, feelings, or intentions. While relating to another person, non-AS's are able to hypothesize more or less what that person is thinking or feeling based on a mental map of their own emotions, and an intuitive knowing of the feelings of other people. Those with AS find it harder to formulate theories or hypotheses about another person's mental or emotional state. Weak Theory of Mind leads to individuals with AS unintentionally and unknowingly saying and doing things in a relationship that can come across as insensitive and be unintentionally hurtful. Over time, the hurt feelings, pain, and suffering of the non-AS spouse can cause some serious tears or lacerations in the marriage.

It is important that both the non-AS and AS spouse become curious and learn about each other's thinking processes, inner worlds, and life experiences, rather than making assumptions or judgments about how the other partner thinks and feels. For meaningful conversation and dialogue to occur, open minds are needed. Verbalizing details about their inner and outer worlds, in a non-judgmental atmosphere, gives partners an opportunity to understand each other better and to bond.

12. Improving communication

Working towards better communication is an ongoing task in any relationship. Within an AS marriage, the importance of communication cannot be stressed enough, since AS is in part characterized as a social-communication deficit. Studies show that 90% of human interaction is based on non-verbal communication. Individuals with AS have difficulties in being able to pick up and interpret facial cues, vocal intonations, and body language, and hence a miss out on a significant amount of communication.

In some cases, the disconnect in an AS marriage is due to the fact that the partner with AS has great difficulty initiating conversations and keeping them flowing. The non-AS spouse feels abandoned and isolated by her AS partner's lack of initiation of connection. The non-AS spouse needs to communicate in clear words everything she would like her AS spouse to know or do on a daily basis. Otherwise, chances are that the AS spouse will not be able to read his partner's mind, due to his somewhat limited Theory of Mind and ability to read non-verbal cues. For both the non-AS partner and the AS partner, verbalizing one's emotional, mental, physical, sexual, spiritual, and social needs in the relationship is the only way to insure that those needs will be met.

The partner with AS is often willing to meet the needs of his partner once he understand exactly what he needs to do. Merely knowing what the non-AS partner's needs *are* is not sufficient for him to know how to meet them. He can, however, learn what to do if he is given concrete, step-by-step actions through which he can offer loving support to his non-AS partner. For example, some spouses may say, "I'm unhappy because we don't talk anymore." It would be more helpful to something like: "I would like for us to have a conversation for about an hour tonight after we put the kids to bed. I'll put the tea kettle on, and then I'd like to tell you about how rough my week at work has been. I don't want you to solve my work problems, I just want you to listen, agree and validate me by saying things like, 'I'm sorry that those things happened. You're brilliant at your job and your company is lucky to have you.'" The more detailed and step-by-step instructions the individual with AS gets, the better he can meet his partner's needs, and the more satisfied she will feel.

13. Co-Parenting Strategies

Individuals with AS can be very good parents when it comes to concrete tasks such as helping the children with their homework, teaching them new skills, playing with them, and taking them on outdoor adventures. When it comes to meeting their children's emotional needs, they might need some coaching and cues from their non-AS partner. The non-AS partner might even have to help their partner with AS to say complimentary things to their children and to schedule one on one quality time with each of the children as well as the entire family on the calendar on a daily and weekly basis. Also, the non-AS parent can help facilitate opportunities for the child to bond with their parent with AS.

Given the complexity and extra challenges of an AS marriage, neuro-diverse couples who do not yet have children may want to think carefully before deciding to become parents. They should assess the strength of their own economic, physical, and emotional resources, and of their additional support networks (extended family, people or services in the wider community). In may neuro-diverse couples, it is probable that the majority of the work of caring for and raising children will fall on the non-AS spouse, as the husband with AS may have executive function difficulties, or may have enough on his plate just managing his other responsibilities, such as holding down a job and keeping himself on an even keel.

14. Managing expectations and suspending judgment

Adjusting one's expectations to accommodate one's partner is important for both the non-AS and the AS partner. Understanding that there are fundamental neurological differences between non-AS's and individuals with AS is important while trying to manage expectations between the partners.

For motivated couples, working hard to improve the marriage with the various tools listed here can bring about real change and made the marriage more comfortable and rewarding for both partners. It is important to note that change and growth is a slow and painful process for any couple or individual wanting to work on their marriage. For any marriage to succeed and thrive long-term, both partners have to make the daily effort to do things differently than they have before. It is also important to understand that growth and change happens in spurts, and that maintaining a high quality and happy marriage is a lifelong commitment.

Couple's counseling for AS marriage

All of the steps and strategies described in this article can be addressed in couple's counseling. With a skilled counselor, experienced in AS, both spouses in the AS marriage will be able to gain awareness of their own individual patterns of behavior, and learn how they can make both attitudinal and behavioral adjustments to get the more out of their relationship. A counselor can also facilitate conversations, and help both partners learn better communication skills. The counselor can also help the couple brainstorm, strategize, connect emotionally, and problem-solve around sensory integration issues, meltdowns, and co-morbid conditions such as anxiety and depression.

If you've met one person with Asperger's, you've met one person with Asperger's.

Chapter 9: Coping with anxiety

Anxiety is a real difficulty for many adults for who have AS. It certainly affects the person psychologically and physically. It can happen for a numerous of reasons, through sensory-based overwhelminess to day to day challenges. We all have our own way of coping with it on a daily basis.

Emotions are somewhat abstract to us but when you are using imagination to decode the specific emotion, it helps to figure out the intensity and the outcome of it. Therefore, you can help yourself to manage it easier.

Emotions are often the inquisitor of physical and mental distortion. They can therefore affect both mind and body inducing several symptoms:

- Losing patience easily
- Difficulty in concentrating
- Thinking constantly about the negative outcome of an action.
- Insomnia
- Depression
- Obsessing and being preoccupied by a certain subject.

There are also physical symptoms that affect us as follows:

- Excessive thirst
- Gastric affections
- Diarrhea
- Aching muscles
- Headaches
- Dizziness

- Tremors
- Pins and needles feeling in arms and legs.
- Nose bleeds
- Urinating frequently
- Tachycardia
- Gas and bloating

If any of these symptoms are experienced by the Aspie or even someone in your surroundings, please be sure to consult your doctor in the near future.

When dealing with anxiety, there are certain strategies that are important to use to overcome them. We must first be able to identify the triggers, learn what causes them and at what frequency.

Keeping a logbook is a great way to follow up with the symptoms and learn to understand them. It helps us remember them also. You can certainly write on how they make you feel and at what moments you experienced them.

Having a meltdown prevention plan is an excellent way to learn to cope with them as they happen and also how to prevent them from happening before hand. An anxiety plan is a list of triggers and situations that causes anxiety but also involves the solutions and strategies that we can use to manage the different levels of anxiety as they arrive. Obviously, the plan can be personalized and adapted depending on how well we understand our anxiety.

Relaxation techniques are also very appreciated by most. People with AS have lots of difficulty in relaxing. Remember that particular hobby or subject of high interest that you may have, you can therefore use them to relax. You can integrate them in your daily routine. Some people with AS need to be left alone to be able to unwind and we must respect that.

Physical activities can help relieve everyday tensions and help expel that stress. Deep breathing exercises help focus and recenter ourselves because we are simply focusing on our breathing. Writing down on a timetable the moments used to relax is a great way to integrate them in your daily routine. Taking a simple bath, listening to relaxation music, aromatherapy in the bedroom or in the sensory room can help reduce the stress we experience.

Talking about our anxiety to a trusted partner can help a great deal as long as the opposite parties are comprehensive and supportive about the matter.

You can also store work or things that are important in a set place, so you will not forget or misplace.

It is really helpful to use the day of the week to help plan and organize tasks, social activities and other events.

Chapter 10: The sensory world of Autism

A lot of people with AS have difficulty processing everyday sensory information such as sounds, sights and smells. This is called having sensory integration difficulties, or sensory sensitivity. It can affect a person's life in great ways.

Our different senses have a special way to work. Our central nervous system processes all our sensory information we receive and helps us organizes it, prioritize and understand the information that we have to process.

Our receptors that are located all over our bodies pick up sensory information or stimuli. Our extremities have the most receptors. We usually process sensory information automatically, without needing to think about the subject. People with AS have sensory integration deficit which have difficulties processing everyday's sensory information around us.

People who have trouble or struggle to deal with all this information at one time may and most likely will become more stressed or anxious and can be accompanied with physical pain. This can challenge behaviors.

We all have five senses: Sight, sound, touch, taste, smell, balance and body awareness are also senses that we have learnt to develop. People with AS can be (over) hypersensitive or (under) hyposensitive in many of these areas.

There are different ways to help People with AS deal with the different senses stimulus. Small changes to their environment can make a very big difference in their lives. These three points are essential when trying to help within these situations of sensory sensitivities.

1. **Be aware**: look all around you to recognize if there is anything that could possibly cause difficulties. Could you change anything in the environment?

2. **Be creative**: Think and create positive and rewarding sensory experiences.
3. **Always be prepared**: Just like the scouts, explain to the person with AS whom you are assisting that there can be possible sensory stimuli's that could be difficult to experience. Create and establish a connection of trust and support.

But how does sensory sensitivity affect their behaviors? In many cases, the person with AS will react in ways that will be hard for you to distinguish and that you would not suspect to be linked to sensory overloads. The person with AS might refuse to wear certain clothes because they don't like the texture or the pressure of that certain piece of clothes. Why not turn the shirt inside out to get rid of the seams, get rid of the itchy tags and why not let them wear clothes that they feel at home in.

There are many more example but they are all different for each individual with AS.

Having a sensory room is a great way to help stimulate, develop and balance the person with AS sensory systems. You can create a sensory room inside your own house. Some people with AS need the opposite, a sensory deprivation room which work the same but while deleting all sensory sensitivities.

A sensory room could include the following :

Soothing music, fiber optics that emit wonderful colors, a beanbag, bubble tubes to make bubbles, vibrating cushions to soothe of stress, putting up pictures in frames that are favored by the Aspie in question.

Just remember to look for the triggers that can set them off and by doing so you can help prevent them before they happen.

Enjoy and Savvy on!

Acknowledgements

I would like to thank my dear husband Normand, without whose help this book would never have been completed.

Thank you for your patience and guidance, your use of the editor's red pen.

www.ingramcontent.com/pod-product-compliance
Lightning Source LLC
Chambersburg PA
CBHW021931170526
45157CB00005B/2282